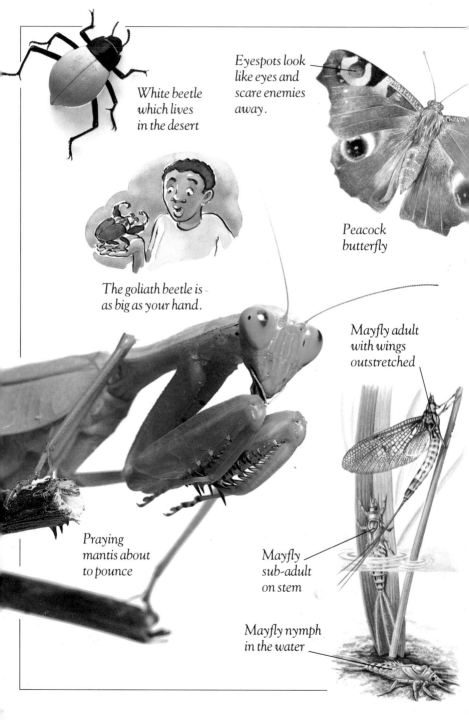

White beetle which lives in the desert

Eyespots look like eyes and scare enemies away.

Peacock butterfly

The goliath beetle is as big as your hand.

Mayfly adult with wings outstretched

Praying mantis about to pounce

Mayfly sub-adult on stem

Mayfly nymph in the water

# Insects

Written by
STEVE PARKER

DORLING KINDERSLEY, INC.

NEW YORK

# A DORLING KINDERSLEY BOOK

**Senior editor** Susan McKeever  **Art editor** Thomas Keenes
**Editor** Jodi Block  **Senior art editor** Jacquie Gulliver
**Production** Catherine Semark  **U. S. editor** Charles A. Wills
**Photography by** Frank Greenaway  **Editorial consultant** John Feltwell

First American edition of this Eyewitness™ Explorers book, 1992
10 9 8 7 6 5
Dorling Kindersley, Inc.,
95 Madison Avenue, New York, NY 10016

Library of Congress Cataloging-in-Publication Data
Parker, Steve.
Insects/ by Steve Parker – 1st American ed.
p. cm. – (Eyewitness Explorers)
Includes index.
Summary: Describes the physical characteristics, behavior, and metamorphosis of insects and
examines kinds of garden insects, woodland insects, insects in the home, and others.
ISBN 1-56458-025-3
ISBN 1-56458-026-1 (lib. bdg.)
1. Insects – Juvenile literature. 2. Insects – Identification – Juvenile literature.
[1. Insects] I. Title. II. Series.
QL467.2.P353 1992                                    91-58212
595.7 – dc20                                              CIP
                                                               AC

Color reproduction by Colourscan, Singapore
Printed in Italy by A. Mondadori Editore, Verona

# Contents

8. Looking at insects

10. What is an insect?

12. Legs and leaping

14. Walking on water

16. On the wing

18. Feeding time

20. Insect-eye view

22. Touch and feeling

24. Taste and smell

26. The mating game

28. Laying eggs

30. Changing shape

32. Growing up gradually

34. Dangerous insects

36. Hunting insects

38. Parasites

40. Living in groups

42. Ant society

44. Busy bees

46. Don't touch!

48. A good coverup

50. Insects in the park

52. Insects in the woodlands

54. Insects in the desert

56. Insects in the water

58. Insects in the tropics

60. Index

# Looking at insects

Many insects are so small that you cannot see them in detail, so they might not seem interesting. But with a few bits of equipment and some practice, you can soon learn about their fascinating ways and habits. Look for insects on leaves and flowers, under stones and logs, and in the soil.

*Pencil-written notes won't run if it rains.*

### Magnifier
A magnifying lens shows an insect in amazing detail. Place it over the insect so it can't run away. But don't keep it trapped for too long!

*Use a notebook with a stiff back.*

### Drawing
Make a sketch of what you see. Label features such as wings or unusually shaped legs. Note the insect's size, color, and where it lives. You can finish the sketch later – and you'll soon improve with practice.

*The common tiger beetle, with its bright green color, is easy to spot as it runs along the sand on sunny days.*

## Make a pooter

A pooter picks up small or fast-moving insects safely. You'll need thin clear plastic, modeling clay, two wide plastic drinking straws, tape, and a bit of muslin.

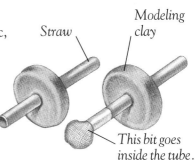

*Straw*

*Modeling clay*

*This bit goes inside the tube.*

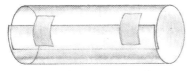

**1** Roll the plastic into a tube 1-2 inches across and tape it. Tape the muslin over the end of one straw.

**2** Make blobs of clay for each end of the tube. Fix a straw in each blob, with the muslin inside the tube.

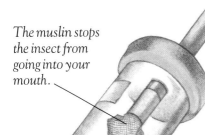

*The muslin stops the insect from going into your mouth.*

**3** Suck the muslin-ended straw. Air rushes in through the other straw, carrying small items with it. Loosen the clay blob to remove things from the tube.

**4** Hold the pooter near an insect on a leaf or a flower and suck.

*Move and lift small insects on a paintbrush to avoid squashing them.*

### Storage jar

A clean glass jar with air holes in the lid makes a good insect home. Always put the insects back outside as soon as you can.

# What is an insect?

There are more kinds of insects than any other kind of animal. There are gnats and midges almost too small to see, irritating flies, stinging wasps, and huge beetles. Check that a creature is an insect by the 3 + 3 rule. A typical insect has three parts to its body and three pairs of legs.

*A butterfly has four large, colorful wings.*

**Ladybug**
This small, tough creature is a kind of insect called a beetle. Under its hard body covering are two wings, ready to fly away if the ladybug meets trouble.

**Butterfly**
The butterfly is a fairly large and delicate insect. Like many other insects, it has two antennae, or "feelers," on its head. These can smell, as well as feel.

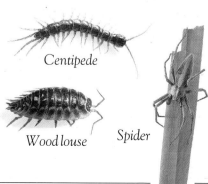

*Centipede*

*Wood louse*     *Spider*

**Not insects**
Count the legs on these animals. Eight legs usually means a spider, and fourteen a wood louse. Lots of legs means a centipede, and even more is a millipede. Only insects have six legs.

## Insect in close-up

The common wasp shows all the main parts of an insect. It has four wings, although many insects have only two. The front and back wings on each side are linked together by hooks, so they flap as one. Wasps have bright yellow and black markings to warn other animals that they are not good to eat. Wasps have a powerful sting, so try not to annoy them!

Feelers

An insect's head has two big eyes, feelers, and mouthparts for biting.

The legs have tiny claws at the end for hanging on to plants.

An insect's leg has joints that bend as it walks and runs.

Wasps have a very small "waist" separating their thorax and abdomen.

The middle part, the thorax, has wings and legs on the outside and large muscles inside.

Tubes called veins stiffen the see-through wings.

The end part, the abdomen, contains the guts for digesting food and the reproductive parts for breeding.

# Legs and leaping

Most adult insects move swiftly on their six legs. They may not seem very fast to us because they are so small. But if an ant was as big as a human, it could run five times faster than an Olympic sprinter. And a human-sized flea could leap over a 40-story building.

**Leg joints**
Like all insects, this bush cricket has six jointed legs, each with four main parts. It also has special "ears" on its front legs for hearing sounds made by other bush crickets.

*Before it leaps, this speckled bush cricket makes sure its feet have a good, firm grip on the branch.*

*The cricket's "ear" is just below the knee joint.*

*The cricket looks around to make sure it is safe to jump.*

**Look before you leap**
Insects with very long legs, such as crickets, grasshoppers, and fleas, are usually good leapers. The cricket folds its long legs back on each other when it gets ready to leap.

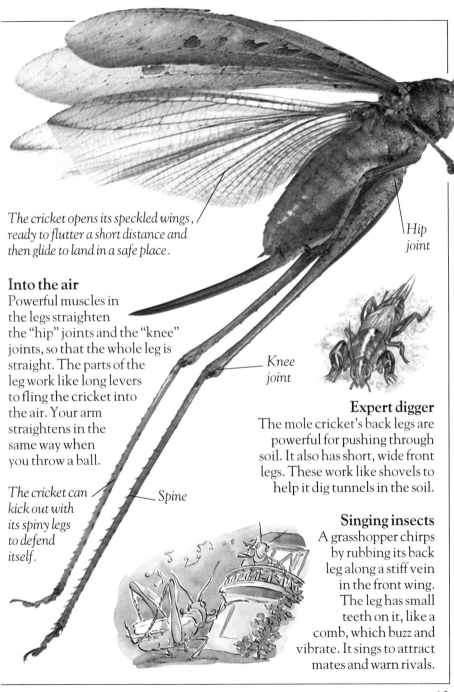

*The cricket opens its speckled wings, ready to flutter a short distance and then glide to land in a safe place.*

## Into the air

Powerful muscles in the legs straighten the "hip" joints and the "knee" joints, so that the whole leg is straight. The parts of the leg work like long levers to fling the cricket into the air. Your arm straightens in the same way when you throw a ball.

*The cricket can kick out with its spiny legs to defend itself.*

Hip joint

Knee joint

Spine

### Expert digger

The mole cricket's back legs are powerful for pushing through soil. It also has short, wide front legs. These work like shovels to help it dig tunnels in the soil.

### Singing insects

A grasshopper chirps by rubbing its back leg along a stiff vein in the front wing. The leg has small teeth on it, like a comb, which buzz and vibrate. It sings to attract mates and warn rivals.

# Walking on water

Where air and water meet, a stretchy "skin" forms at the water's surface. Because many insects are tiny, this skin is strong enough to support them. Water striders and whirligig beetles can run on it. Underwater insects hang from it and poke their breathing tubes up into the air, or collect air bubbles.

*Rear legs are rudders for steering. Middle legs push the strider across the surface. Front legs detect ripples from struggling prey.*

**Water boatman**
Look out for the water boatman swimming just under the surface of rivers, lakes, and ponds. It rows along on its back, using its large rear legs. It hunts tiny fish, tadpoles, and insects. Don't touch – it can give a painful bite!

**Surface skater**
Water striders skim over the water surface. Waterproof hairs on their feet make dimples in the water as they chase after insects trapped in the surface skin.

**Whirligigs**
These small, black beetles swim in groups, spinning on the water surface. They eat dead creatures that fall into the pond.

*The whirligig beetle has two pairs of eyes for seeing above and below the surface.*

## Make an insect aquarium

Study insects from ponds or streams in a small aquarium for a week or two. Catch them by dipping with a net and put them back afterward. Which insects live near the surface or on the bottom? When are they most active?

*Many pond insects can fly. Keep a lid with a close-weave net or mesh on the tank.*

*Molting insects such as mayflies need plants or sticks above the surface to crawl on.*

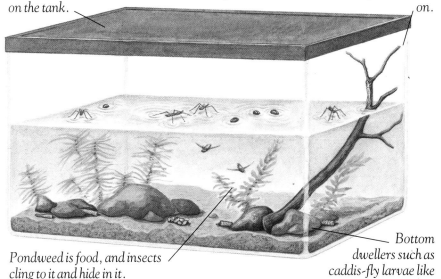

*Bottom dwellers such as caddis-fly larvae like to hide under stones.*

*Pondweed is food, and insects cling to it and hide in it.*

**1** A glass fish tank makes a good insect aquarium, but any large bowl will do. Put gravel and small stones on the bottom. Add pond or fresh water and plants to make the tank about half full. Let it settle for a few days and then add your insects.

*Change two jelly jars full of tank water for pond water each day to keep the water fresh and to provide food.*

**2** Keep the tank in a well lit but cool corner of the room. The water will get too hot in bright sunlight.

**3** Tiny plants called algae may grow as a green scum on the sides of the tank. Rub them off with a sponge tied to a stick, and filter them out of the water with a strainer.

# On the wing

Most insects have wings and can fly at some stage in their lives. Insect wings are thin and light, and are stiffened by a network of tubes called veins. Strong muscles in the middle section of the insect provide the power for flapping.

**Lacy wings**
The lacewing's beautiful pattern of veins in its four wings looks like finely woven lace. This insect flies mainly at night.

**Beetle take off**
Beetles have four wings. But the front pair is hard and tough. They cover the beetle's back and protect the long, folded flying wings underneath. This cardinal beetle is preparing for takeoff.

*The cardinal beetle lifts its red protective wing covers.*

*It unfolds and spreads out its main pair of flying wings.*

## Up and away

The long, see-through wings, with their strengthening veins, move so fast that they look like a blur. The beetle flies off to look for more food.

*Main wings do all the flapping.*

*The hard wing cases do not flap at all. The beetle holds them clear of the main flying wings.*

## Speeding up

With each beat, the main wings push air downward. This lifts the cardinal beetle upward. Then the wings also begin to push air backward, so that the beetle moves forward.

*The beetle beats its wings faster and faster until they reach flying speed. Pushing with its legs against the flower, it springs away. In a flash, the beetle is airborne.*

*The beetle straightens its legs and holds them down, away from the fast-moving wings.*

## Expert flier

Hoverflies are expert fliers. They can hover above flowers and then dart off and reappear a short distance away. They can move in any direction – including backward.

17

# Feeding time

Some insects can eat almost anything. Cockroaches have strong chewing mouthparts to eat stale food, dirt and even paper and leather. Other insects have specially shaped mouths. Bugs have a hollow, needle-shaped beak to pierce and suck sap from plants. Look at plant leaves for evidence of hungry insects.

## Nipping weevil
Weevils are a kind of beetle. They have a long, curved snout called a rostrum, with tiny jaws at the end. They eat small pieces of leaves and other plant parts. If you look on a nettle plant, you may spot a weevil.

*The curved fang is called a mandible.*

*The long, tube-shaped mouth is called a proboscis.*

## Spearing prey
This diving beetle larva has large, pointed fangs for spearing prey. It squirts special juices into the tadpole that dissolve its body into soup. The beetle larva then sucks up the mushy meal.

*This is a hummingbird hawkmoth – it feeds like a hummingbird.*

## Built-in straw
Butterflies and moths have mouths shaped like drinking straws. When they are not feeding, the straw is coiled up. To suck sweet nectar from flowers, it is uncoiled into a straight tube.

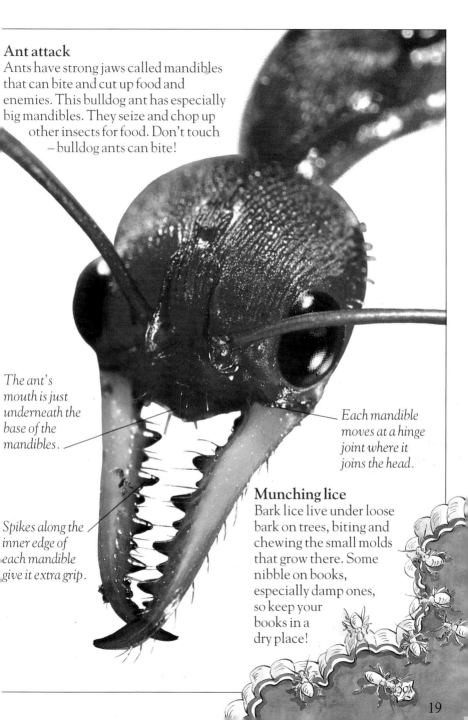

## Ant attack
Ants have strong jaws called mandibles that can bite and cut up food and enemies. This bulldog ant has especially big mandibles. They seize and chop up other insects for food. Don't touch – bulldog ants can bite!

*The ant's mouth is just underneath the base of the mandibles.*

*Each mandible moves at a hinge joint where it joins the head.*

*Spikes along the inner edge of each mandible give it extra grip.*

## Munching lice
Bark lice live under loose bark on trees, biting and chewing the small molds that grow there. Some nibble on books, especially damp ones, so keep your books in a dry place!

19

# Insect-eye view

What does an insect see? Probably a very different view of the world than we see. Most insects have eyes made of many separate sections, called ommatidia (*om-a-tid-ee-a*). Each section sees a small part of the surroundings. The whole view may look like a patchwork.

### Simple eyes

Besides two main eyes (compound eyes), many insects also have several smaller eyes. These are known as simple eyes or ocelli (say *o-sell-ee*), and they do not have separate sections. They probably just detect blurs of light or dark.

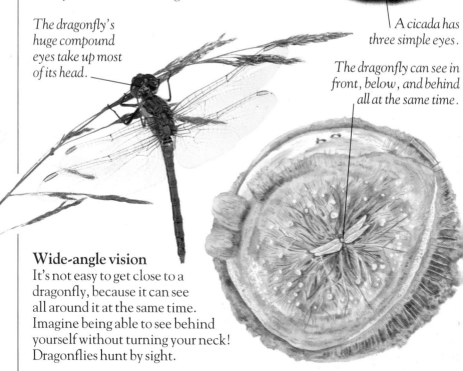

*The dragonfly's huge compound eyes take up most of its head.*

*A cicada has three simple eyes.*

*The dragonfly can see in front, below, and behind all at the same time.*

### Wide-angle vision

It's not easy to get close to a dragonfly, because it can see all around it at the same time. Imagine being able to see behind yourself without turning your neck! Dragonflies hunt by sight.

*Each ommatidium has a transparent top to let in light.*

*Light is refracted, or broken up, into rainbow patterns off the tiny ommatidia.*

## What big eyes . . .

Dragonflies have the biggest eyes of any insect. This is a front view of the common darter dragonfly shown opposite. Each eye has up to 5,000 ommatidia! Many insects can tell colors apart, and many, such as moths, can see clearly at night.

*The eyes look for danger, mates, rivals, and egg-laying places.*

## Seeing the invisible

Many insects can detect ultraviolet light. This is light from the sun that we cannot see (but it gives us a tan). Guidelines on petals show up only in ultraviolet light. Honeybees see them and follow them to the nectar.

# Touch and feeling

A good sense of touch helps insects to detect air movements when flying and to cling onto leaves and stems. Insects feel for footholds when crawling, for food when feeding, and for holes and cracks when hiding.

*Antennae detect scents, tastes, and currents of air.*

*Just sitting on a leaf, the chafer beetle keeps its antennae closed.*

### Branching out

Insect antennae are very sensitive to touch. The chafer beetle has branched, fan-shaped antennae. Normally, it keeps them closed. When it takes off, the beetle fans out the antennae. This makes it more sensitive to wind currents and any smells in the air.

*When in flight, it fans its antennae out.*

### Extra-long antennae

The long-horned moth's thin antennae are more than twice as long as its body. They bend and wave using even thinner muscles inside them. Big groups of moths dance around trees on sunny days. Their antennae stop them from bumping into one another.

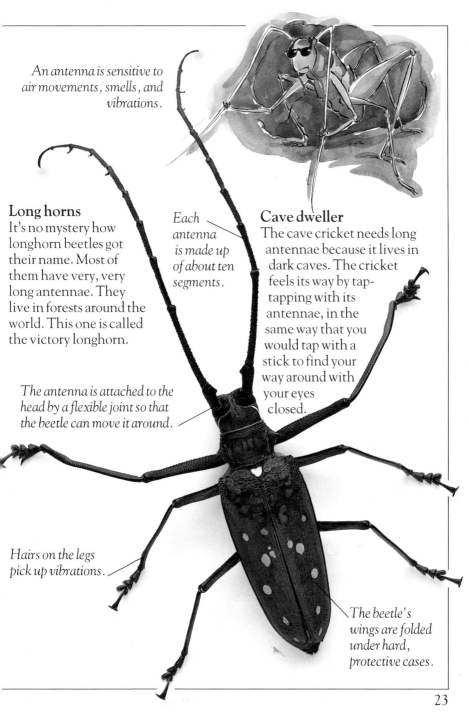

An antenna is sensitive to air movements, smells, and vibrations.

## Long horns

It's no mystery how longhorn beetles got their name. Most of them have very, very long antennae. They live in forests around the world. This one is called the victory longhorn.

Each antenna is made up of about ten segments.

The antenna is attached to the head by a flexible joint so that the beetle can move it around.

## Cave dweller

The cave cricket needs long antennae because it lives in dark caves. The cricket feels its way by tap-tapping with its antennae, in the same way that you would tap with a stick to find your way around with your eyes closed.

Hairs on the legs pick up vibrations.

The beetle's wings are folded under hard, protective cases.

# Taste and smell

Insects do not have a nose for smelling and a tongue for tasting as we do. Instead, they have tiny taste buds on many parts of the body, mainly on their mouthparts and antennae but also on their feet! The main parts for detecting smells floating in the air are the antennae.

*The bluebottle is a common sight in kitchens.*

**Fussy flies**
Tasting and smelling usually help an insect to find either food or a mate. This bluebottle fly smells the syrup, which could be food. It lands and checks by tasting to see if it is good to eat.

*The fly spreads digestive juices on the food, which is like vomiting over it!*

*The fly has taste buds on its feet. These are the first parts to touch the food .*

*Fly dabs up dissolved food with its sponge-like mouthparts.*

## Smelly meal
Look out for dungflies buzzing around dung. They are attracted by the smell of fresh animal droppings! They lay their eggs in dung. When the maggots hatch, they eat the droppings, which contain lots of nourishment.

## Follow the smell
When a searching ant finds plentiful food, it lays a trail of invisible scent on the ground, leading back to the nest. The other ants follow the trail and find the food easily.

## Testing tastes and smells
Try a simple test to see which insects are attracted to each dish by smell. When they land and taste it, do they stay or fly off? Do you think they can detect pure water? Can you?

**1** Put syrup or sugary water into one small dish, a bit of old meat or gravy into another, and tap water into the third.

**2** On a sunny summer day, put the dishes outside on a table, two to three feet apart. Note which insects visit them in your notebook.

**3** Butterflies and wasps smell the sweet syrup, which they eat as food. Flies smell the meat and come to lay eggs, since their maggots eat flesh.

25

# The mating game

Most insects need to mate in order to make more insects. The female must mate with a male of her own kind before she can lay eggs. Usually, the male seeks out and courts the female. Courtship involves dancing, tapping the body, releasing a smell, and even flashing lights or fighting your partner!

*The glow comes from the underside of the female's body.*

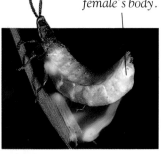

## Lighting-up time
The glowworm is not a worm; it's a type of beetle. At night, the female makes a yellowish light. This attracts a male glowworm that mates with her.

### Perfumed chamber
The male bark beetle builds a special mating chamber. Then he produces a tempting smell that floats into the air. When the female bark beetle detects it, she can't help but follow the smell trail to him.

*Male mantis has had its head eaten off.*

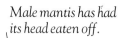

### Meal for a mate
The male mantis must be very careful as he comes near the larger female to mate. If she is hungry, she may eat him!

## Mating bugs

These shieldbugs are mating. The male passes his sperm from an opening in his rear end into the female's opening. The sperm join up with the eggs in the female's body. When she lays the eggs, they are ready to develop into young shieldbugs.

*Both partners keep still while mating, so they are at less risk from predators.*

*The sperm pass from the male shieldbug to the female.*

*If you see insects joined together like this on a leaf or a twig, leave them alone.*

*The green shieldbug has a very strong covering that protects its body like a warrior's shield.*

*The shieldbugs hold on tight to the leaf.*

27

# Laying eggs

Each kind of insect finds the best place to lay its eggs. In the wrong place, the eggs could dry out, get too wet, turn moldy, or be eaten. Also, the young insects need food nearby when they hatch. A dungfly lays eggs in fresh droppings, because this is what the maggots eat when they hatch!

*The female checks that the plant is the right one before she lays.*

## Life underneath a leaf

The underside of a groundsel, or ragwort, leaf is the ideal place to look for the shiny yellow eggs of a cinnabar moth. The females lay eggs there because the caterpillars (developing moths) eat the same leaves when they hatch. This is called laying eggs on the right food plant.

*She lays about 30 eggs on the leaf's underside, hidden from most egg-eaters.*

*The wood wasp taps and prods the wood with her ovipositor, feeling for the best place to drill.*

## Drilling deep

Wood wasps lay their eggs through a tube called the ovipositor. In some insects, such as this ichneumon (*ik-new-men*) wasp, the ovipositor is like a long needle. It drills deep under the wood's surface, and the eggs slip out of its tip.

## Underwater eggs

Insects such as damselflies and dragonflies lay their eggs in or near water because the young live in water after hatching. The female damselfly pushes her thin abdomen under the surface and glues her eggs to the plant stems below.

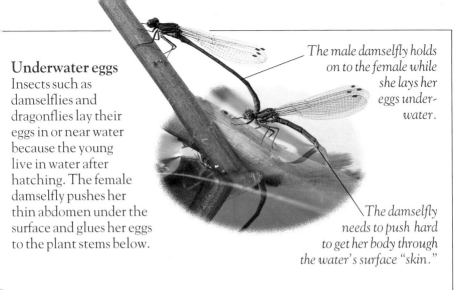

*The male damselfly holds on to the female while she lays her eggs underwater.*

*The damselfly needs to push hard to get her body through the water's surface "skin."*

## Eggs in a purse

To protect her eggs from danger, the cockroach lays them into a bag or purse called an ootheca. She produces this smooth, tough purse from her abdomen.

*The leathery bag contains two rows of eggs.*

## Home sweet home

The female cockroach carries the purse around for a few days. Then she tucks it into a safe home, such as a warm corner or crack, and glues it firmly in position. The young develop and hatch two months later, as worm-like larvae.

# Changing shape

*The worm-like stage of the beetle's life is called a larva.*

Many newly hatched insects look nothing like the adults they become. Young flies, beetles, butterflies, bees, ants, and wasps all look a bit like worms. They grow bigger, then change their body shape completely and turn into adults. This way of growing up is called complete metamorphosis.

*As the larva grows, it must split its skin and wriggle out.*

*New, bigger skin forms underneath. Changing skins like this is called molting.*

### The larva
A newly hatched larva of a flour beetle is a real insect, but it looks like a small worm. It is called a mealworm because it eats "meal," an old name for ground grains and flours, like oatmeal.

*Chrysalis hangs on a twig.*

### Becoming a butterfly
Butterfly eggs hatch into larvae called caterpillars. Each outgrows its skin and becomes a pupa – known as a chrysalis (*kris-a-lis*) in butterflies. Chrysalises are found hanging from branches or on the underside of leaves. Many look brown and wrinkled, like old leaves. Inside its case the caterpillar is changing shape. In a few weeks, it becomes a butterfly.

## The pupa

The larva continues to eat, grow, and molt its skin. After about five molts, it turns into the next stage of its life, a hard-cased pupa. The pupa does not move at all, but inside, the larva is changing enormously.

*Antenna*

*Mouthparts*

*Claws on feet*

*Pupa*

*Inside the pupa, parts of the larva's body break down, then rebuild into a beetle.*

## The adult

After a few weeks, the pupa's case splits open and an adult flour beetle crawls out. Only now does it look like a typical insect, with six legs and wings.

*Wings are under the wing case on its back.*

## Wings out to dry

When an adult insect like this cicada comes out of the pupa, its wings are all crumpled. The adult pumps blood into them. The wings spread out and become stiff and hard as they dry. Then the adult can fly away.

# Growing up gradually

Some insects don't go through as many stages of growth as others. The young of these insects hatch from their eggs looking like tiny adults. These babies are called nymphs. These nymphs molt their outer skins many times, changing body size slightly at each molt. This is called incomplete metamorphosis. Insects that grow like this include locusts, dragonflies, and bugs.

### Hatching
The female locust lays her eggs in a strong case in the sand. The baby locusts hatch from the eggs and dig their way out.

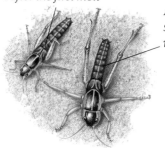

*After the first molt*

*After the second molt*

*The nymph's claws hold it firmly on the twig.*

### Getting bigger
Every few weeks, the locust nymph molts and then grows before the new skin hardens. It does this four times. It only has small wings and cannot fly. But at the fifth molt, an adult emerges from the skin, complete with full-length wings.

*Adult begins to wriggle out of the old skin.*

*Adult's body enlarges as it pulls its long legs out from inside the old nymph legs.*

## Not quite adult

A mayfly nymph lives for years underwater, breathing with three tail-like gills. At last, it crawls up a plant stem and splits its skin. But it does not become an adult yet. First, it molts into a sub-adult, which cannot fly very well. The sub-adult flutters to a bush or tree near the water. In a few hours, it sheds a thin skin and becomes the real adult, which can fly much better.

### First flight

As soon as its wings have dried, the adult locust can fly away.

*The old skin is now almost empty.*

*Adult hangs onto nymph skin . . .*

*The empty nymph skin is perfect in every detail, even down to the antennae and eye coverings.*

*. . .as it pulls the rest of its body free.*

*The adult's wings begin to spread out as blood pumps into their veins.*

# Dangerous insects

Most of the millions of kinds of insects are not harmful, but a few cause problems. Some caterpillars, beetles, and weevils spoil crops and damage trees. Cockroaches infest food supplies. Termites weaken wooden furniture and buildings. Some insects kill people by biting and spreading diseases.

Abdomen full of blood

### Jungle fever

Different kinds of mosquitoes spread the germs that cause malaria and yellow fever. These illnesses occur in warm tropical places such as swamps and jungles, where mosquitoes live and breed.

### Spreading germs

When a mosquito bites someone and sucks blood, it takes in any germs in the blood. When it bites another person, some of the germs leak into the blood. This may spread the germs that cause diseases.

### Beware of the sting

Insects such as bees and wasps have a sharp stinger at the rear end. A bee can use its stinger only once, but wasps can use theirs several times.

You can recognize the hornet by its yellow and reddish-brown coloring.

The stinger pokes out of the rear end as the hornet attacks.

The hornet, a big wasp, can give a very painful sting – so be careful!

## Potato pest

Colorado beetles love to eat potato leaves. If they breed well, they cause enormous damage to potato crops. Originally, these beetles came from the Rocky Mountains in Colorado – which explains their name.

*The aphids collect on buds where the sap is easiest to get to.*

## The gardener's enemy

Look closer at a flower, and you may see this sight – lots of little insects crawling around. These are aphids, also called greenflies and blackflies. Aphids use their spear-like mouths to pierce plants and drink the sap. They breed very fast. When the young hatch, they feed on the leaves and soon the plant shrivels and dies. Aphids feed on roses, beans, cabbages, and other plants.

*A newly born aphid, or nymph, is the same shape as its mother.*

## Quick babies

In early spring, a female aphid does not need to mate or lay eggs. She gives birth to babies, that quickly grow and produce their own young. This is why aphids breed so fast.

# Hunting insects

Insects that hunt other creatures need to have large eyes to see their prey, strong limbs to hold it, and powerful jaws to crunch and chew it. Some have stingers to poison and paralyze their victims. The hunter on this page is called a praying mantis. It munches on other insects, such as flies – and also other mantises.

**How to catch a fly**
The mantis keeps perfectly still as it watches a fly. Its green color makes it look like a harmless green leaf. Even its eyes are green!

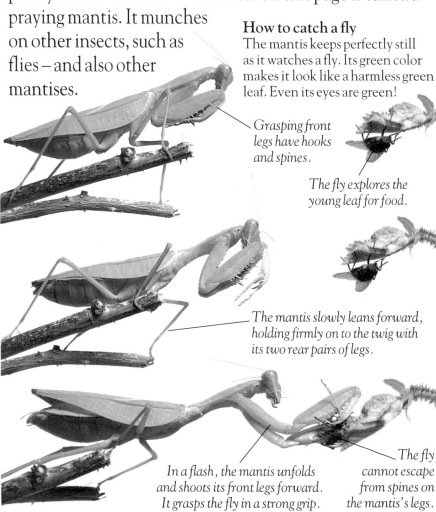

*Grasping front legs have hooks and spines.*

*The fly explores the young leaf for food.*

*The mantis slowly leans forward, holding firmly on to the twig with its two rear pairs of legs.*

*In a flash, the mantis unfolds and shoots its front legs forward. It grasps the fly in a strong grip.*

*The fly cannot escape from spines on the mantis's legs.*

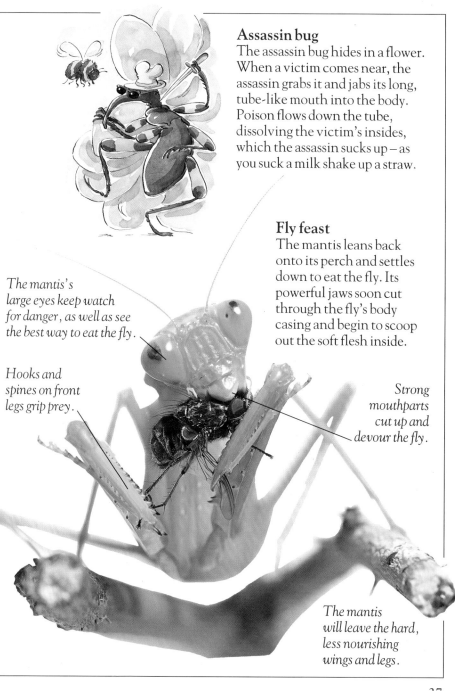

## Assassin bug

The assassin bug hides in a flower. When a victim comes near, the assassin grabs it and jabs its long, tube-like mouth into the body. Poison flows down the tube, dissolving the victim's insides, which the assassin sucks up – as you suck a milk shake up a straw.

## Fly feast

The mantis leans back onto its perch and settles down to eat the fly. Its powerful jaws soon cut through the fly's body casing and begin to scoop out the soft flesh inside.

*The mantis's large eyes keep watch for danger, as well as see the best way to eat the fly.*

*Hooks and spines on front legs grip prey.*

*Strong mouthparts cut up and devour the fly.*

*The mantis will leave the hard, less nourishing wings and legs.*

37

# Parasites

A parasite is an animal (or plant) that lives on or in another, which is known as the host. The parasite often harms the host, usually by stealing its blood or other nourishment. Several kinds of insects are parasites. Lice, mosquitoes, and fleas suck blood. Many kinds of wasps are parasites on the eggs, larvae, or adults of other insects.

**Weevil hunter**

The weevil-hunting wasp chooses big weevils as hosts for its larvae. As the weevil feeds on a flower, the wasp grasps and stings it. The sting paralyzes the weevil so it cannot move – but it does not die.

*The wasp holds the weevil in position with its strong legs.*

*The stinger enters where the weevil's body casing is thin, usually at a joint.*

*A flea can leap more than a foot with its strong rear legs.*

### Leaping bloodsucker
Worm-shaped flea larvae live in dirt. They turn into adults that pierce the skin of their host and suck its blood. Different types of fleas have different hosts. Fleas from cats, dogs, and rats sometimes bite people.

### Irritating itch
If a flea bites you, it leaves a red spot on your skin that will itch for a few days. Rat fleas spread diseases such as the plague called the Black Death, which killed millions of people in the Middle Ages.

*The weevil can't move.*

### Food for the young
The wasp takes several weevils to her nest in a burrow she has dug in sandy soil. She lays her eggs on the weevils and then fills in the burrow. When the larvae hatch, they eat the live hosts.

*Weevils in burrow wait to be eaten.*

### Weevil breakfast
The wasp larvae eat away their hosts' flesh. After a few weeks, the weevils die. The larvae turn into pupae, and then into adults that dig their way out.

# Living in groups

Have you ever noticed that some insects are never alone? These are called social insects – they need to live in a group in order to survive. Termites are social insects that look and live like ants. They make huge nests, with one queen ruling many workers. Some feed on fungi.

Queen

Workers are blind and wingless.

Soldier

King

### Egg machine
The queen termite's body is huge, filled with eggs. She mates with the smaller king, and then lays 30,000 eggs each day! She is fed, cleaned, and cared for by workers. Soldier termites with strong biting jaws defend her and the nest.

### Mending nests
Many animals, including birds, feed on termites. When they try to break into the nest, workers gather to repair the hole. These termites are mending their nest in a tree trunk.

### A new life
In a big termite colony, some young termites grow up with eyes and wings. In suitable weather, they fly away to find a mate and start a new nest.

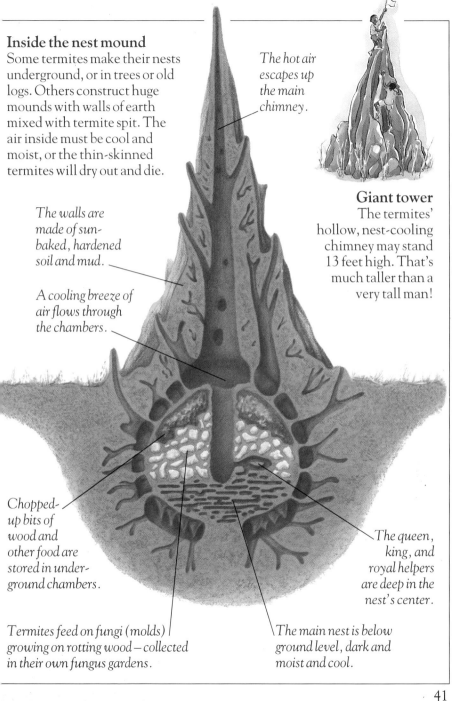

## Inside the nest mound

Some termites make their nests underground, or in trees or old logs. Others construct huge mounds with walls of earth mixed with termite spit. The air inside must be cool and moist, or the thin-skinned termites will dry out and die.

*The hot air escapes up the main chimney.*

### Giant tower

The termites' hollow, nest-cooling chimney may stand 13 feet high. That's much taller than a very tall man!

*The walls are made of sun-baked, hardened soil and mud.*

*A cooling breeze of air flows through the chambers.*

*Chopped-up bits of wood and other food are stored in underground chambers.*

*The queen, king, and royal helpers are deep in the nest's center.*

*Termites feed on fungi (molds) growing on rotting wood – collected in their own fungus gardens.*

*The main nest is below ground level, dark and moist and cool.*

# Ant society

Scurrying ants are social insects, like termites. The queen ant is the only one that lays eggs. The workers look after her and feed her. They go in search of food, including plants, seeds, and even other insects. Big-jawed soldiers defend the underground nest.

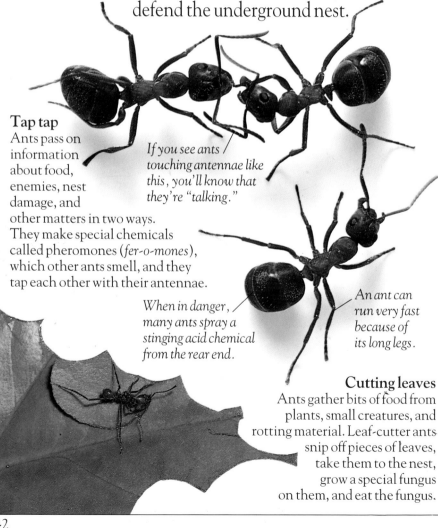

**Tap tap**
Ants pass on information about food, enemies, nest damage, and other matters in two ways. They make special chemicals called pheromones (*fer-o-mones*), which other ants smell, and they tap each other with their antennae.

*If you see ants touching antennae like this, you'll know that they're "talking."*

*When in danger, many ants spray a stinging acid chemical from the rear end.*

*An ant can run very fast because of its long legs.*

**Cutting leaves**
Ants gather bits of food from plants, small creatures, and rotting material. Leaf-cutter ants snip off pieces of leaves, take them to the nest, grow a special fungus on them, and eat the fungus.

## Make an ant farm

Borrow ants from nature and watch them for a few days to see them working together, making a nest, and gathering food. Use an empty aquarium tank, old goldfish bowl, or other ant-tight container.

1 Stick dark paper around part of the tank. The ants like their tunnels to be dark, so they are more likely to make them up against the sides in these parts.

2 Look for an ants' nest in the yard, park, or meadow. Under a stone is a good place. Gather a few trowel-fuls of ants and soil. Try to get a mixture of different - sized ants.

3 Put the ants and soil in the tank. Add more damp (not wet) soil, a few new and old leaves, and a bit of fruit. Make sure air can get in but ants can't get out.

4 After a few days, gently lift the dark paper. You may see ant tunnels built up against the glass. Put the ants back where you found them after a week or two.

# Busy bees

You may have seen honeybees buzzing from flower to flower. This means that they're collecting food to bring back to their nest or hive (a house built for them by people). Like termites and ants, honeybees live in groups.

*The bee sucks nectar with its tube-shaped mouthparts.*

🖐 *Bees can sting – so watch their antics from a safe distance.*

*It carries the food back to the hive.*

## Bee line

When a worker bee finds some flowers that have lots of sweet nectar to collect, it flies back to the hive and "dances" to tell the other workers where it is. Then they fly hundreds of times between flowers and hive.

## Bags of pollen

As the bee sips nectar, it rubs against pollen grains (yellow powder on flowers). The pollen gets caught on the bee's body. The bee carries the pollen back to the hive in hairy "shopping bags" on its back legs.

## Keeping bees

People keep honeybees so that they can collect their sweet, sticky honey. The bees change the nectar into honey, which they store in honey cells. The beekeeper removes some combs and takes the honey.

# Don't touch!

If you notice a brightly colored insect in the park or wood – you are probably meant to! The bright reds, yellows, and oranges, often patterned as spots or stripes, are warning colors. They warn predators such as birds and lizards that the insect has horrible-tasting flesh, that it bites or stings, or that it squirts foul-smelling fluid.

*Froghoppers can hop well, like frogs.*

### Blister bearer
The blister beetle's shiny body warns that this beast tastes bad. If its blood gets on the skin of a person or animal, it causes painful blisters. Animals soon learn to leave the beetles alone.

*The froghopper's pattern is like the ladybug's, which has black spots on a red body.*

*Blister beetles gather on flowers to feed and lay eggs.*

### I taste horrible
Bright red spots on the black body of the red-and-black froghopper mean it tastes horrible. Young froghoppers make a ball of frothy spit to cover them up as they suck plant juices.

## Inside the hive

A big hive may have more than 50,000 bees. The queen lays all the eggs. Lots of female workers feed and clean her, clean the hive, sting enemies, and look after the eggs and the larvae in their cells. There are also males, called drones. Several drones mate with the queen.

*A worker dances by waggling her abdomen to tell other bees where the nectar is.*

*Workers make six-sided cells from wax, which comes from glands below their abdomens.*

*Some cells will contain stored honey. Others will hold eggs and growing larvae.*

45

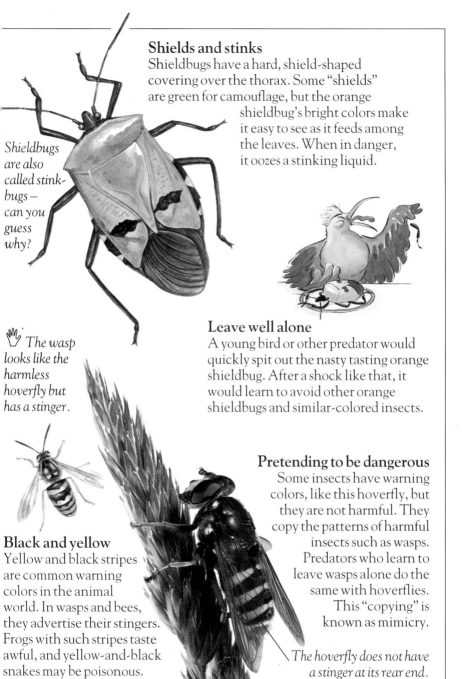

## Shields and stinks

Shieldbugs have a hard, shield-shaped covering over the thorax. Some "shields" are green for camouflage, but the orange shieldbug's bright colors make it easy to see as it feeds among the leaves. When in danger, it oozes a stinking liquid.

*Shieldbugs are also called stink-bugs – can you guess why?*

*The wasp looks like the harmless hoverfly but has a stinger.*

## Leave well alone

A young bird or other predator would quickly spit out the nasty tasting orange shieldbug. After a shock like that, it would learn to avoid other orange shieldbugs and similar-colored insects.

## Pretending to be dangerous

Some insects have warning colors, like this hoverfly, but they are not harmful. They copy the patterns of harmful insects such as wasps. Predators who learn to leave wasps alone do the same with hoverflies. This "copying" is known as mimicry.

*The hoverfly does not have a stinger at its rear end.*

## Black and yellow

Yellow and black stripes are common warning colors in the animal world. In wasps and bees, they advertise their stingers. Frogs with such stripes taste awful, and yellow-and-black snakes may be poisonous.

47

# A good coverup

Some insects look like insects. Others don't. They look like bright green leaves, old brown leaves, colorful petals, twigs, buds, thorns, or even birds' droppings. Such a disguise helps the insect to stay unnoticed by predators. Some insects frighten off predators with marks like huge, fierce eyes.

### Prickly problem

Thornbugs suck the sap of plants such as roses and brambles. Their bodies are shaped like the plant's own thorns and prickles, which few predators want to eat!

*Thornbugs have to stay still for their disguise to work.*

*Eyespots look like those on a peacock's tail. This is why it's called the peacock butterfly.*

### Butterfly or hawk?

With wings folded over its back in its resting position, the peacock butterfly looks like a piece of bark. But when in danger, it opens its wings and flashes huge eyespots, pretending to be a hawk.

*The dark markings on the underside of the wings help it to blend in with the bark.*

*Thorns are like the insect's spines.*

*Claws on the feet grip the twig securely, even in a strong wind.*

*The flattened bulges on the legs look like the blades of old, shrunken leaves.*

*The bent abdomen has spiny edges, like the leaves and twigs around it.*

### Spiny stick

This giant spiny stick insect, up to 8 inches long, blends in perfectly with the prickles and curly leaves on which it's perched. Predators ignore it, as long as it stays put on the prickly plant.

### Cricket camouflage

The bright color of the green bush cricket is like the green of the leaf it sits on. Blending in with the surroundings like this is called camouflage.

# Insects in the park

Even the neatest, best-kept park or garden is home to thousands of insects. In the summer, you can see them crawling in the soil, climbing up plants, eating leaves, and buzzing or flitting from flower to flower. If it gets cold, they hide in cracks in walls or under roofs.

*The devil's coach horse beetle defends itself by arching its tail and squirting horrible-smelling liquid from its rear end.*

### Careful mother

In springtime, earwigs lay their pearly, oval-shaped eggs. The mother earwig is one of the few insects that cares for her eggs. She carefully licks them all over to keep them from becoming moldy.

### Nighttime hunter

You won't see the devil's coach-horse beetle by day unless you look in damp places under stones, leaves, or logs. It comes out at night to catch tiny caterpillars, worms, and spiders.

*The bumblebee sucks nectar up through its tube-like mouthparts.*

## Helpful bumble

Insects such as bumblebees visit flowers to collect pollen and nectar. They help the flower by carrying its pollen to other flowers of the same kind so they can develop into seeds.

*The bumblebee brushes pollen onto its back legs to take back to its nest.*

## Borrowing insects

To study insects, borrow them from nature using a pitfall trap. You'll need a jelly jar, a trowel, and a flat stone. Afterward, put them back in the same place. Are the same insects active by day as at night?

**1** In a sheltered corner, dig a small hole in the ground, just big enough for the jelly jar. Put some old leaves in the jar. Find four stones.

**2** Put the jelly jar in the hole, its rim at ground level. Place the flat stone over it, propped up on the smaller stones, to keep out rain.

**3** During the day, look in the jar every two or three hours. Note which insects have fallen in. Remove the jar when you've finished.

# Insects in the woodlands

Woodlands are ideal places for insects. The trees provide food and shelter for hundreds of different kinds of insects. Leaves, buds, fruits, and seeds are all food for plant-eating insects. Hunting insects prey on the plant-eaters. Some insects burrow into living wood or eat mold growing under loose bark. Others feed deep in the roots or rotting wood.

*The beetle's huge, antler-like jaws are not strong enough to bite you.*

### Stag beetle
A stag beetle larva looks like a big, white worm. For three years, it lives in an old woodland log or stump, eating the wood. Then it becomes a pupa, and finally an adult. Only males have the huge jaws, which look like deer antlers.

*The stag beetle has wings under its body casing. You might see one flying on a warm summer night.*

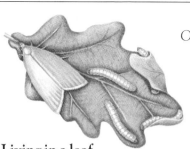

## Living in a leaf
Green-oak roller moths are found wherever oak trees grow. The young moths (caterpillars) roll up oak leaves and eat them by the thousands!

*Oak-apple galls, made by tiny wasp larvae, look like small green apples that slowly turn brown.*

*Gall wasp – ten times life size*

## Tree-dwellers
To see tree-dwelling insects more clearly, put a white sheet under a low branch. Knock the base of the branch with a stick (but do not break the branch). Caterpillars, beetles, crickets, and other insects may fall onto the sheet.

## Gall-makers
Some tiny wasps, flies, and other insects live, feed, and lay eggs inside leaves, twigs, and buds. The tree grows a tough wall around the eggs and seals them off. These growths are called galls. They come in many strange shapes. Look out for lumps on twigs and leaves.

## Locking horns
The male stag beetle uses its giant jaws to fight off rival males. Sometimes males lock jaws and have a pushing contest. But the jaw muscles are so weak that they cannot harm each other.

*The female stag beetle's jaws are small. But she can give a painful bite.*

# Insects in the desert

Deserts are too dry for most animals. But many insects can live there, especially beetles. They usually hide by day to avoid the hot sun. At night they feed on plants, wind-blown seeds, or one another. Some burrow into desert plants such as cacti. Others feed on the bodies of creatures that died from heat or thirst.

**Sandy beetle**
This desert darkling beetle is sand-colored to hide it from enemies. Its larvae look like worms or maggots and burrow into the sandy soil, as earthworms do in garden soil.

**Sand trap**
The ant lion larva, also called a doodlebug, digs a cone-shaped pit in the desert sand and hides at the bottom. When an ant tumbles in, the larva grabs it and eats it.

**Adult ant lion**
Adult ant lions look a bit like lacewings. The males of some kinds dance in midair in swarms to attract females at breeding time.

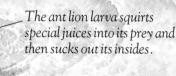

*The ant lion larva squirts special juices into its prey and then sucks out its insides.*

54

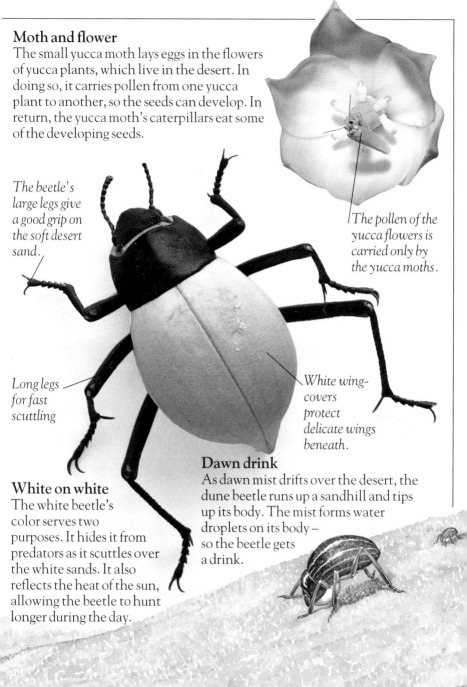

## Moth and flower

The small yucca moth lays eggs in the flowers of yucca plants, which live in the desert. In doing so, it carries pollen from one yucca plant to another, so the seeds can develop. In return, the yucca moth's caterpillars eat some of the developing seeds.

*The beetle's large legs give a good grip on the soft desert sand.*

*The pollen of the yucca flowers is carried only by the yucca moths.*

*Long legs for fast scuttling*

*White wing-covers protect delicate wings beneath.*

## White on white

The white beetle's color serves two purposes. It hides it from predators as it scuttles over the white sands. It also reflects the heat of the sun, allowing the beetle to hunt longer during the day.

## Dawn drink

As dawn mist drifts over the desert, the dune beetle runs up a sandhill and tips up its body. The mist forms water droplets on its body – so the beetle gets a drink.

# Insects in the water

Next time you're passing a pond or river, take a closer look at the water, for many insects make their homes in water. There are insects that skim over the surface and ones that swim upside down. And there are some insects, like dragonflies, that live in water only when they are young.

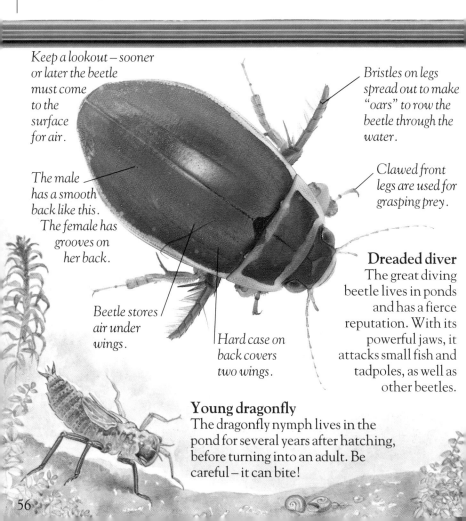

Keep a lookout – sooner or later the beetle must come to the surface for air.

Bristles on legs spread out to make "oars" to row the beetle through the water.

The male has a smooth back like this. The female has grooves on her back.

Clawed front legs are used for grasping prey.

Beetle stores air under wings.

Hard case on back covers two wings.

**Dreaded diver**
The great diving beetle lives in ponds and has a fierce reputation. With its powerful jaws, it attacks small fish and tadpoles, as well as other beetles.

**Young dragonfly**
The dragonfly nymph lives in the pond for several years after hatching, before turning into an adult. Be careful – it can bite!

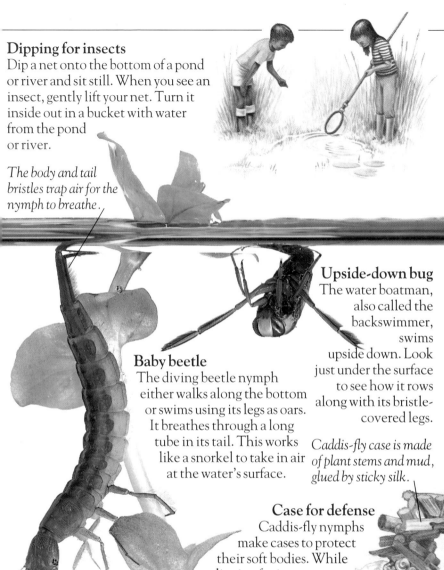

## Dipping for insects

Dip a net onto the bottom of a pond or river and sit still. When you see an insect, gently lift your net. Turn it inside out in a bucket with water from the pond or river.

*The body and tail bristles trap air for the nymph to breathe.*

### Upside-down bug

The water boatman, also called the backswimmer, swims upside down. Look just under the surface to see how it rows along with its bristle-covered legs.

### Baby beetle

The diving beetle nymph either walks along the bottom or swims using its legs as oars. It breathes through a long tube in its tail. This works like a snorkel to take in air at the water's surface.

*Caddis-fly case is made of plant stems and mud, glued by sticky silk.*

### Case for defense

Caddis-fly nymphs make cases to protect their soft bodies. While dipping for insects, you may find stones, stems, or shells stuck together. If you look closely, you may see a caddis-fly nymph in its case.

*Huge, sharp mouthparts spear tadpoles, small frogs, and fish.*

57

# Insects in the tropics

Millions of different kinds of insects live in the jungle. Insects love the warmth and dampness. There are plenty of plants to feed on and hide in, and masses of tiny animals to catch. You may often see the largest, strangest, and most colorful insects in these places.

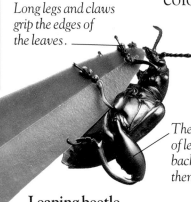

*Long legs and claws grip the edges of the leaves.*

*The rear pair of legs fold back on themselves.*

### Leaping beetle

The back legs of the frog beetle are long and strong, like a frog's legs. It uses them to leap away from predators. The beetle then opens its wings and flies away to complete its escape.

### Double disguise

The pink-flower mantis looks just like a beautiful flower and is found on orchids. When small creatures come near to search for pollen or nectar, the mantis grabs them in its fearsome front legs. This special disguise also protects the mantis from creatures that might eat it, such as birds and lizards.

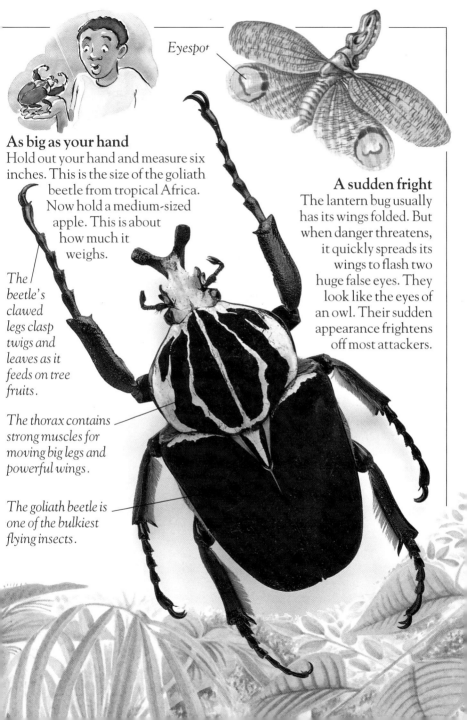

*Eyespot*

## As big as your hand
Hold out your hand and measure six inches. This is the size of the goliath beetle from tropical Africa. Now hold a medium-sized apple. This is about how much it weighs.

*The beetle's clawed legs clasp twigs and leaves as it feeds on tree fruits.*

*The thorax contains strong muscles for moving big legs and powerful wings.*

*The goliath beetle is one of the bulkiest flying insects.*

## A sudden fright
The lantern bug usually has its wings folded. But when danger threatens, it quickly spreads its wings to flash two huge false eyes. They look like the eyes of an owl. Their sudden appearance frightens off most attackers.

# Index

## A

abdomen, 11, 29, 45
ant, 12, 19, 25, 30, 40, 42-43, 44
antennae, 10, 11, 22-23, 24, 31, 33, 42
ant lion, 54
aphids, 35

## B

backswimmer, 57
bee, 30, 34, 44-45, 47
beetle, 10, 16, 18, 30, 34, 38, 53, 54;
  bark, 26;
  blister, 46;
  cardinal, 16-17;

*Goliath beetle*

  chafer, 22;
  Colorado, 35;
  darkling, 54;
  devil's coach horse, 50;
  dune, 55;
  flour, 30-31;
  frog, 58;
  goliath, 59;
  great diving, 18, 56-57;
  longhorn, 23;
  stag, 52-53;
  tiger, 8;

whirligig, 14;
  white, 55
breeding, 11, 26-29, 35, 54
bug, 18, 32;
  assassin, 37;
  lantern, 59
bumblebee, 51
butterfly, 10, 18, 25, 30;
  peacock, 48

*Grasshopper*

## C

caddis-fly, 15, 57
camouflage, 47, 49
caterpillar, 28, 30, 34, 50, 53, 55
centipede, 10
chrysalis, 30
cicada, 20, 31
cockroach, 18, 29, 34
colors, 8, 11, 34, 36, 46-49, 55
cricket, 53;
  bush, 12-13, 49;
  cave, 23;
  mole, 13

*Dragonfly*

## D

damselfly, 29
doodlebug, 54
dragonfly, 20-21, 29, 32, 56;
  darter, 21
dungfly, 25, 28

## E

ears, 12
earwig, 50
eyes, 11, 14, 20-21, 33, 36
eyespots, 48, 59

## F

feelers, *see* antennae
feet, 14, 24, 31, 49
flea, 12, 38, 39
flight, 16-17
fly, 10, 25, 30, 36, 37, 53;
  bluebottle, 24
froghoppers, 46

## GH    *Waterstrider*

galls, 53;
  oak-apple, 53
glowworm, 26
gnat, 10
grasshopper, 12, 13

hornet, 34
hoverfly, 17, 47

## L

lacewing, 16, 54
ladybug, 10, 46
larva, 18, 29, 30-31, 38, 39, 45, 52, 53, 54
lice, 38;
  bark, 19;
  book, 19
locust, 32-33

# M

maggots, 25, 28, 54
malaria, 34
mantis, 26;
  flower, 58;
  praying, 36-37
mayfly, 15, 33
mealworm, 30
metamorphosis, 30-33
midge, 10
millipede, 10
mimicry, 47
molting, 15, 30, 31, 32

mosquito, 34, 38
moth, 18;
  cinnabar, 28;
  green- oak roller, 53;
  hummingbird hawk-, 18;
  long-horned, 22;
  yucca, 55
mouthparts, 11, 18-19, 24, 31,
  37, 44, 51, 57

# NO

nymph, 32-33, 35, 56, 57

ommatidia, 20, 21
ovipositor, 28

# P

parasitic insects, 38-39
pheromones, 42
pitfall trap, 51
pooter, 9
proboscis, 18
pupa, 30-31, 39, 52

# RS

rostrum, 18

shieldbug, 27, 47
singing, 13
social insects, 40-45
spider, 10, 50
stick insect, 49
stings, 10, 11, 34,
  36, 44, 45,
  46, 47

*Cricket*

# T

taste buds, 24
termite, 34, 40-41, 44
thorax, 11
thornbug, 48

# W

wasp, 10, 11, 25, 30, 34, 38,
  47;
  gall, 53;

*Colorado beetles*

  ichneumon, 28;
  weevil-hunting, 38-39
water boatman, 14, 57
water strider, 14
weevil, 18, 34, 38, 39
wood louse, 10

# Acknowledgments

**Dorling Kindersley
would like to thank:**
Sharon Grant and Carol
Orbel for design assistance.
Gin von Noorden and Kate
Raworth for editorial
assistance and research.
Jane Parker for the index.

**Illustrations by:**
Julie Anderson, John Davis,
Nick Hewetson, Mark Iley,
Simon Thomas.

**Picture credits**
t=top b=bottom c=center
l=left r=right
Bruce Coleman Ltd: 20t,
26tl; /Jane Burton 12t, 44b,
54tl; /G.Cubitt 40b; /
M.Fogden 42bl, 46tr, 55tr; /
Kim Taylor 40c; /Peter Ward
25tl.
Oxford Scientific Films: /
Kathie Atkinson 19c; /J.A.C.
Cooke 34cl; /Stephen Dalton
14b; /L.S.F. 29c.
Premaphotos: /Preston-
Mafham 18tl, 49bl.